NARROW GAUGE NET

NET

UMMER SPECIAL

CW00551877

WELCOME!

Greetings! And welcome to this, the second *Narrow Gauge Net Summer Special*! We were very pleased with the feedback which last summer's edition received from both readers, reviewers and trade customers.

There has been no change in the format of the *Special*, although we have tweaked the content slightly, acting on feedback received from our readers.

Once again, we are endeavouring to showcase the best of the UK narrow gauge scene, with a couple of overseas interludes to add variety. We hope that you enjoy it - your continued contributions to either the *Summer Special* or the monthly e-zine would of course be most welcome!

Iain McCall
Editor

CONTENTS

Our grateful thanks to all who have contributed to this "Summer Special", and to those who have taken the time to read it! Contributions for future issues will be gratefully received.

COVER PICTURES

Front Cover: A line which is taking great steps to improve its enthusiast appeal and restore its Great Western character and heritage is the Vale of Rheidol. This view shows No. 9 in Cambrian Railways livery basking in the sun at Devil's Bridge on 10th June 2014.

Rear Cover: A thoroughly atmospheric view of PRINCE departing across the Cob with the Vintage Train during the FR's Victorian Weekend on 12th October 2014. The rearmost vehicle is the distinctive Curly Roofed Van.

Title Page: The fireman of Quarry Hunslet UNA opens the injectors during a layover at Cei Llydan on the Llanberis Lake Railway during their Gala event on 5th July 2014.

All Cover Pictures by Peter Donovan

Published by Mainline & Maritime Ltd
3 Broadleaze, Upper Seagry, near Chippenham, SN15 5EY
Tel: 01275 845012
www.mainlineandmaritime.co.uk
Trade and Retail Enquiries welcome!

Printed in the UK by Contract Publishing UK
www.contractpublishinguk.co.uk

ISBN: 978-1-900340-35-9

THAT WAS THE EVENT THAT WAS!

No Motor Rails were harmed in the taking of this photograph! Three years of planning and preparation came together spectacularly in the award-winning "Tracks to the Trenches" event at the Moseley Railway Trust's Apedale Valley Light Railway in September, with the above "accident" one of many photographic cameos staged, to the delight of both the railway enthusiasts, the military re-enactors, and the general public. The concensus from the re-enactors the photographer spoke to was that nobody will better this event during the whole five years of WW1 commemorations!

Iain McCall

More active cameos were taken by the various Motor Rails participating, including LR3090, above, hauling a WDLR H Class tank wagon; right, LR2573 propelling a stretcher bearer through the encampment; and, below right, LR 2478 providing a resting platform for the French Army, whilst the British Army continue to work its predecessor in the background! The photo below shows two of the multinational steam contingent present, in the form of Apedale resident JOFFRE (representing France), and (representing the USA) visiting Baldwin 778 from Leighton Buzzard. There were numerous representatives of the British and Commonwealth steam fleets in attendance, along with a Feldbahn making a guest appearance on behalf of the enemy!

Peter Donovan, Keith Frewin (3)

THE SCOTTISH NARROW GAUGE
by David Scotney

Scottish narrow gauge in action. A 1970 Andrew Barclay diesel-mechanical locomotive (ex ICI (Nobel) Works at Ardeer and Powfoot) waits to haul a train at Livingston Mill Station on the 2ft 6in gauge Almond Valley Railway in 2014. Other stock, including a Baguley-Drewry locomotive, can be seen in a siding.

Courtesy of the Almond Valley Heritage Trust

So what is the narrow gauge heritage of Scotland? It is generally not the popular bucolic image of rural narrow gauge railways – for Scotland the narrow gauge railway mainly had a somewhat more dour workaday aspect! However some of the remaining examples in Scotland do now have somewhat the more bucolic feel.

As an aside, it may well be worth mentioning that a number of early Scottish lines opened between 1826 and 1840, amounting to over 150km of route were built to the marginally 'sub-standard gauges' of 4ft 6in or 4ft 6½in but were subsequently regauged to standard as the national network consolidated.

Left: Reconstruction of a station on the pre-1970s Subway at the Glasgow Riverside Museum of Transport.

Ed Webster

Right: Hillhead Station today.

Courtesy of SPT

Passenger Lines

Scottish narrow gauge passenger lines were predominantly urban based. The most well-known narrow gauge railway is the Glasgow Subway. The 4ft (1.219m) gauge Subway was an early underground railway, opened in 1896 by the private Glasgow District Subway, with operation using cable traction. Following financial problems the Subway was sold to Glasgow Corporation in 1922. The Corporation electrified the Subway in the early 1930s with the new operations starting, utilising converted original rolling-stock, in 1935. A display relating to the Subway in this period may be found at the Glasgow Riverside Museum of Transport (www.glasgowlife.org.uk/museums/riverside/Pages/default.aspx). In 1973 the Greater Glasgow Passenger Transport Executive (GGPTE) was established, and took over the Subway from the Corporation. The GGPTE proceeded to totally modernise the system with new rolling-stock, and it reopened (after a 2 year closure) in 1979. The successor to GGPTE, the Strathclyde Partnership for Transport (SPT), is now undertaking a further total modernisation of the Subway with the support of the Scottish Government. The Subway, which now carries around 13 million passengers per year (from a peak of 37 million in 1950), will in the future be operated with new driverless trains. For anyone travelling on the Subway for the first time the interior size of the vehicles does give a somewhat claustrophobic feel when compared with the 'tubes' in London; although the nominal diameter of the tunnels is 11ft (3.253m) compared with the original tunnel diameter of the standard gauge City & South London Railway of 10ft 2in (3.1m - although these tunnels have since been increased to the 'London-standard' of 11ft 8in (3.56m)). With these small tunnels the trains do feel 'narrow gauge' and indeed this must be the most heavily trafficked narrow gauge railway in the UK.

There were also seven Scottish narrow gauge urban electric tramways: six of which utilised the tramway '*UK narrow gauge standard*' of 3ft 6in gauge (1.067m - Dunfermline & District (29.5km), Kirkcaldy Corporation (9.8km), Perth Corporation (8.0km), Rothesay & Ettrick Bay Light Railway (7.8km), Wemyss & District (11.9km) and the Great North of Scotland Railway line serving its Cruden Bay Hotel (1.1km)); and Falkirk & District (12.5km) which used the somewhat strange tramway gauge of 4ft (1.219m). While most of these served predominantly urban routes, some lines at Cruden Bay, Dunfermline, Rothesay, and Wemyss operated through quite open countryside. Most tramways served a predominantly local demand; however

Formed from components of Nos 1 & 2, this is the preserved Cruden Bay Hotel tramcar at the Grampian Transport Museum.

Iain McCall

the Rothesay & Ettrick Bay Light Railway also had a very strong tripper trade from summer patrons who had travelled '*doon the watter*' on the Clyde steamers from the Glasgow area, while the Cruden Bay line was focussed purely on Hotel residents and supplies. All these lines were opened for electric operation between 1899 and 1909 and closed in the 1930s.

A slight anomaly, which is also worth noting, is that the enormous Glasgow tramway system (over 200 km) was also 'just' narrow gauge – it used 4ft 7¾in (1.416m) gauge to allow standard gauge railway wagons to run over its tracks to and from various shipyards along the Clyde (the difference was due to the differing wheel profiles on railways and tramways). The Glasgow trams lasted until 1962, the last complete system to close in the British Isles, and examples of the cars can also be found in the Glasgow Riverside Museum of Transport and at Summerlee Museum of Scottish Industrial Life.

The only 'conventional' public railway was the 5 mile 2ft 3in gauge Campbeltown & Machrihanish Light Railway running across the narrow Kintyre peninsula. This line had been opened to serve a colliery in 1877, but introduced a passenger service in 1907 after gaining a Light Railway Order. The canny financial concept, developed by the colliery engineer, was to cater for the large number of holidaymakers in the area (carried once again '*doon the watter*' by the Clyde steamers) during the summer when the colliery line had very little coal traffic. The railway had three small four-coupled tank locomotives prior to the introduction of passenger services, but to cater for the increased traffic two 0-6-2Ts were bought in 1906/7 from Andrew Barclay. In addition four long (43ft 6in – 13.26m) bogie open-platform passenger carriages were acquired from R Y Pickering & Co of Wishaw. This stock catered for demand until, unfortunately after the First World War, bus competition appeared at the same time as coal production was falling; coal transport finally finished in 1929 and the line did not survive for the 1932 summer passenger season.

A postcard 'flyer' to advertise the 2ft 3in gauge Campbeltown & Machrihanish Railway which was distributed on the Clyde steamers.

Industrial Lines

There have been at least 125 separate locomotive-hauled narrow gauge industrial lines in Scotland; spanning from short 2ft gauge lines for peat harvesting to a 25 mile 3ft gauge line for the maintenance of a hydro-electric water tunnel. The industries that they have served have included: brickworks, construction, defence establishments, explosive works, gas works, holiday camps, hydro-electricity plants, mines, peat workings, quarries, refuse disposal, sawmills, steelworks, and timber extraction. They have also served private shooting estates and serviced private houses.

However if a few primarily Scottish narrow gauge themes can be picked out for these lines they would include:

- The shale oil industry;
- The explosives / defence industries; and
- Local authority services.

The Shale Oil Industry

Oil-bearing shale can be found in the Central Belt of Scotland, primarily within the area of the present West Lothian. Oil from the shale had been extracted and utilised over the years; but the larger scale industrial exploitation did not start until Dr James 'Paraffin' Young set up a business refining oil from colliery workings in 1848. Young's business gradually grew through the 1850s and other entrepreneurs joined the industry from the 1860s. By 1873 524,000 tons of shale were mined annually and this grew to 3,280,000 tons by 1913. Employment in the industry at its peak was some 12,000 people and it supplied around 2% of world oil production. The heyday of the industry lasted from about 1893 to 1925. However competition in oil production from the USA and elsewhere reduced potential demand and price. In 1919 the remaining works merged to form Scottish Oils Ltd, which became a subsidiary of the Anglo-Persian Oil Company (later renamed the British Petroleum Company Ltd (BP) in 1954). The industry survived thereafter by a blend of the closure of smaller operations and government support until the final plants ceased in 1963.

The shale was mined from narrow seams in rock, accessed from both drift and vertical shafts. Most shale mines were considered to have relatively low levels of

A 2ft gauge 1936 10hp Ruston & Hornsby in Westwood Shale Pit.
Ray Hooley Ruston & Hornsby Archive

A general view of the approach to the Oil Works at Winchburgh. Shale-filled tubs are waiting on the left to be rope-hauled up the incline in the background to the top of the retorts. On the right empty tubs and a passenger train, headed by a 1940s Andrew Barclay locomotive, wait to return to the mines.
SRPS

explosive gas concentrations compared with those in coal mines (although there were indeed gas explosions on a number of occasions).

Once extracted the shale was transported to processing plants where it was crushed into uniform pieces, dropped into a vertical retort within which it slowly travelled down through increasing temperatures to produce a mix of crude oil and ammoniacal liquor together with spent shale, the spent shale was deposited onto waste bings, while the liquor was separated into its constituent parts. The crude oil was then refined in a two stage distilling process to produce various different types of oil which could then be processed in a variety of ways to provide usable products.

The huge bings are one of the most obvious remains of the industry, with their reddish colour from the high temperature retorting process, although the spent shale from many of them has been reused in recent years as construction fill for roads etc.

Railways were used within the mines to transport the shale, then to move the shale from the mines to the crushers and onto the retorts, to carry the spent shale from the retorts to the bings, to transport the crude oil from the retorts to central refineries, and then to distribute the final products. The role of narrow gauge railways was in the early parts of the process within the mines and from there to the retorts and the bings.

Within the mines 2ft was the predominant gauge, there is evidence of its use at Burngrange Pits, Fraser Pit, Philpstoun Mines (Linlithgow), Roman Camp Mines (Broxburn), Uphall Mines and Westwood Pit. 2ft 6in was used at Hopetoun Mine. Haulage in the mines was undertaken initially by hand or pony. In the 1930s however light diesel locomotives were developed considerably in efficiency and for use in mines in terms of both 'exhaust conditioning' (the removal of harmful or irritant components from the exhaust gases) and 'flameproofing' (the removal of any possibility of ignition by the locomotive (from sparks, high temperatures, etc.) of explosive gases in the environment). The leading UK manufacturers at this time were Ruston & Hornsby and Hunslet, with both introducing 'exhaust conditioning' in 1936 and 'flameproofing' in 1939. With their lower gas concentrations shale mines were able to introduce diesels with only 'exhaust conditioning' and they were introduced from 1936 at Westwood Pit. Some 17 Ruston & Hornsby diesels were delivered between 1936 and 1949 to all the 2ft gauge pits mentioned above. Hopetoun Mine unusually utilised battery locomotives.

The distance from the mines to the tops of the retorts was usually quite short and in a number of cases continuous rope haulage was utilised for the narrow gauge tubs. This was also used in most cases to haul the hot spent shale from the foot of the retorts to the top of the bings, although the movement of the tubs on the top of the bings to the actual tipping areas was generally man-hauled.

There was one example of a specific narrow gauge railway running from mines to a major retorting centre: the 2.5 mile (4km) 2ft 6in gauge Winchburgh Railway linking Duddingston Nos. 1, 2, 3 & 4 Mines, Philpstoun No.6 Mine, Whitequarries Mine and Totleywells Mine southward to the Niddry Castle Oil Works at Winchburgh. This was opened in 1902 by the Oakbank Oil Company and at that stage only linked to the Duddingston Mines. It was unusually electrified with an overhead supply at 500V DC. In later years it was cut back at the north end from Duddingston Nos.1 & 2 Mine, which closed in 1941, but was linked westwards with Philpstoun No.6 and Whitequarries Mines from a junction near Duddingston Nos. 3 & 4 by an endless rope operated 2ft 6in line, and a new electrified link ran eastwards (south of Duddingston No.3 & 4 Mine) to serve Tottleywell Mine. While the railway primarily carried the shale from the mines, it also provided an internal passenger service between the mines and the workers village at Winchburgh. The two initial open 4-wheel 50hp locomotives came from Baldwin in the USA with Westinghouse electrical equipment. These were reinforced in 1907 with a 100hp 4-wheeler from British Westinghouse (using Bagnall mechanical components). No.4, another 100hp machine, arrived in 1929 from English Electric but this time with a central cab. The final two 60hp/72hp locomotives came in 1942 and 1946 from Andrew Barclay using Metropolitan Vickers electrical equipment. The line remained in operation until February 1961, after which the track was quickly removed and five of the locomotives scrapped. The remaining locomotive, Baldwin No.2 of 1902, owned by the National Museums of Scotland is at the Almond Valley Heritage Centre.

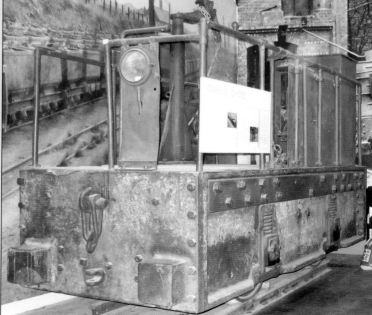

The 1902 Baldwin/ Westinghouse locomotive from the Winchburgh Railway, owned by the National Museums of Scotland, is now at the Almond Valley Heritage Centre.
Author

A 2ft 6in gauge 4-wheeled Hunslet (works No.8967 of 1980) shunting explosives wagons at the Royal Ordnance Factory at Bishopton on 29th August 1991.

Courtesy of Gordon Edgar

The Explosives and Defence Industries

The wide open spaces of Scotland and their proximity to the sea provided potentially relatively safe locations for the manufacture, stock-piling and transport of explosives, particularly for the defence industries.

Nobel Industries was founded by Alfred Nobel in 1871 for the production of dynamite. Later the company diversified into many other types of explosives. The first UK factory was established at Ardeer on the Ayrshire coast, with its design based on other Nobel sites in Sweden and elsewhere. By 1907 it was reputed to be the largest explosives factory in the world, at its peak it covered 8 square kilometres and employed 13,000 people. In 1926 the company merged with others to form Imperial Chemical Industries (ICI), with Ardeer continuing to operate under the ICI (Nobel Division). (http://www.nobelprize.org/alfred_nobel/biographical/articles/dolan/). In 2002 ICI sold Ardeer and after a major fire in 2007 the site was closed down. Another smaller ICI (Nobel) site was to be found at Powfoot near Annan which seems to have existed from the 1940s to 1990s.

The Government also, inevitably, was involved in the manufacture of explosives, shells etc.. The largest of these sites in Scotland (and the UK) was the Royal Ordnance Factory (ROF) at Bishopton near Paisley. A factory was established here in the First World War, and by 1917 was employing 10,000 workers. It closed between the Wars, but was revived and reconstructed at the start of the Second World War. By 1941 it had a workforce of 20,000 and covered an area of 8 square kilometres. It was sold in the 1980s to Royal Ordnance Plc and then to British Aerospace. Production at the site ceased in 2002.

Government storage / stockpiling of munitions was undertaken at a number of sites across Scotland including Crombie in Fife (Naval Ordnance Depot from 1915) and Eastriggs near Gretna (Central Ammunition Depot from the 1930s using a 5 square kilometre part of the site of the First World War Gretna Munitions Factory).

The Crombie Naval Ordnance Depot was opened in 1915 to support the new naval dockyard at Rosyth, and was equipped from the start with a 2ft 6in internal rail network like those at the existing naval sites at Priddy's Hard and Bedenham (Portsmouth, Hampshire) and Chattenden/Lodge Hill (Kent). Similar 2ft 6in networks for the navy were subsequently developed at new munition storage sites in the 1920s at Ernesettle (Plymouth, Devon); and in the 1930s at Broughton Moor (Cumbria), Trecwn (Dyfed) and Dean Hill (Wiltshire). The 2ft 6in network at Crombie closed in the 1980s although the site remains in use.

The commercial sites at Ardeer and Powfoot appear to have been equipped with comprehensive 2ft 6in internal networks in the 1930s and 1940s respectively. The networks closed with the sites.

Eastriggs and Bishopton were both equipped with extensive internal narrow gauge networks as part of their reconstruction in the 1930s. That at Eastriggs used 2ft gauge, which matched similar sites developed in the same period by the Ministry of Defence (MoD - army and RAF) elsewhere in the UK such as Chilmark (RAF, Wiltshire), Corsham (CAD, Wiltshire), Fauld (RAF, Staffordshire), and Monkton Farleigh (CAD, Wiltshire). Bishopton used 2ft 6in gauge, as was also provided at its contemporary ROF at Bridgewater / Puriton (Somerset). The Bishopton network closed with the site, and the Eastriggs site (and network) have been gradually abandoned over about the last 10 years.

Obviously rail haulage within explosives factories and munition storage sites had to use methods which could not trigger an explosion. Initially therefore at Crombie four small fireless steam locomotives from Andrew Barclay (which specialised in this

The Scapa Flow Museum situated in the old naval base at Lyness, in the Orkney Isles, includes in its display a small collection of narrow gauge artefacts, pertaining to the use of the narrow gauge in and around naval bases. Most of the NG exhibits, including the BE seen here, are stored overnight in a container, and positioned by fork lift truck at the start of each opening day.

Iain McCall

A 1980 2ft 6in gauge 'flameproofed' Baguley-Drewry, originally delivered to the Navy at Trecwn, but then transferred to Dean Hill, and now at the Almond Valley Railway.
Author

type of locomotive) were used, which had their steam reservoirs filled from a central boiler facility away from the munitions area, and these remained in operation until the Second World War. Later small battery locomotives were used at Bishopton (from 1939) and Crombie (from 1943) built initially by Wingrove & Roberts and Greenwood & Batley; and later (1970 onwards at Bishopton) by Brook Victor. In the 1940s Crombie had the four fireless and about seven battery locomotives; while the huge Bishopton network had about 40 battery locomotives.

As had been noted above, the 1930s saw the introduction of light diesel locomotives which could operate safely in the potentially explosive environment to be found in mines. This form of haulage now therefore also offered a safe option for explosives and munitions sites. The Ardeer and Powfoot systems were equipped with new Ruston & Hornsby diesels from the start; while Eastriggs, after the initial temporary use of an array of secondhand contractors' diesels, obtained a mix of new Ruston & Hornsby and Hunslet diesels. During the Second World War Ardeer had eight and Powfoot two diesel locomotives; while Eastriggs had about 12 diesel locomotives.

It may be noted that many of the locomotives attached to Crombie were moved around between this site and other naval munition storage sites, and similarly those at Eastriggs circulated with other MoD sites.

After the War there was some re-equipping of the various sites: Ardeer and Powfoot had new Ruston Hornsbys in the 1950s and Andrew Barclays, MotorRails and Hunslets in the 1970s; Eastriggs obtained new Hunslets in the 1970s and had Baguley-Drewrys of the 1970s transferred from Chilmark after its closure in the 1990s; and somewhat surprisingly Bishopton took new Hunslets in the 1970s and 1980s and MotorRails in the 1980s.

Of these various sites, only Crombie remains on active military service and none retain any operating narrow gauge railway equipment.

Local Authority Services

The Glasgow Subway, and trams, have already been mentioned; but Glasgow Corporation was also one of the largest users of narrow gauge railways in Scotland in its extensive 'empire' of other essential services. It operated Gas Works (until taken over by the Scottish Gas Board in 1949) at: Dalmarnock was served by an internal 2ft gauge network (5 Kerr Stuart/Bagnall 0-4-0Ts 1914-1956); Dawsholm also with a 2ft gauge network (12 Sharp Stewart/Drummond/Kerr Stuart/Bagnall 0-4-0Ts 1893-1964); Provan with a 2ft 6in network (9 Barclay 0-4-0Ts 1903-1958); and Tradeston with a 2ft network (9 Sharp Stewart/Kerr Stuart/Bagnall 0-4-0Ts 1912-1953). In addition the Corporation also used 2ft 6in gauge at its Govan Refuse Disposal Works (3 Hudswell Clarke battery-electrics 1932-1961), 2ft gauge at Palace Rigg peat works (1 Lister petrol mechanical tractor 1930-1946) and also 2ft gauge at various housing construction sites between 1935-1950 (44 MotorRail diesel-mechanical tractors). One 1946-built Andrew Barclay (Kilmarnock) 0-4-0T locomotive from the Provan Gas Works still survives as No.8 DOUGAL on the Welshpool and Llanfair Railway in Wales.

Other municipal enterprises across Scotland, although on a considerably smaller scale than Glasgow, also used narrow gauge lines. Examples included: gas works for Dundee Corporation (2ft, four Kerr Stuart 0-4-0Ts, 1900-1950s) and Edinburgh Corporation (2ft, four Andrew Barclay 0-4-0Ts, 1903-1960s) until these were also taken over by Scottish Gas Board in 1949; as well as road stone quarries for Banff County Council (2ft, two diesel mechanical tractors, 1950s/60s), Lanarkshire County Council (2ft gauge, four petrol tractors, 1940s-1960s), and Roxburgh County Council (2ft, three petrol-mechanical tractors, 1930s-1960s).

Diminutive 2ft 6in gauge 1946 Andrew Barclay 0-4-0T, designed for working in the restricted surroundings of Glasgow Corporation Provan Gas Works, now named DOUGAL at the Welshpool & Llanfair Railway.

'Fairlightworks'

Locomotive Construction

The other area relating to narrow gauge railways in Scotland was the large concentration of locomotive builders. These included many well-kent company names which provided the motive power on narrow gauge, and other, railways across the former British political and economic empires. The main firms were: the mighty North British Locomotive Company based in Glasgow (with its predecessors Dubs & Co, Neilson & Co and Sharp Stewart & Co); and the smaller Kilmarnock firm of Andrew Barclay & Sons. In addition it should also not be forgotten that the firm of Alley & MacLellan of Glasgow established the 'Sentinel' name relating to steam wagons, but passed on the allied construction of railway products (many narrow gauge) to its subsidiary, and later separate company, in Shrewsbury. Narrow gauge reminders of the Scottish locomotive-building industry can be found at three main sites:

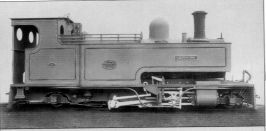

One of the North British Locomotive Company works photographs to be found at the Mitchell Library in Glasgow: a 1903-built 2-6-2T (Works No.16058) for the 2ft 6in gauge Delhi-Umballa-Kalka Railway Company (Kalka – Simla Railway) in India, which was the prototype for a series of 30 similar locomotives built by North British between 1904 and 1910 (and 10 more built in the 1930s and 1950s by Henschel and Krauss-Maffei in Germany).
Courtesy of ©CSG CIC Glasgow Museums and Libraries Collection: The Mitchell Library, Special Collections

- A visit to the Mitchell Library in Glasgow rewards with an amazing array of the original works photographs of many of the 26,000 locomotives built by the North British Locomotive Company and its predecessors (http://www.glasgowlife.org.uk/libraries/the-mitchell-library/special-collections/north-british-locomotive/Pages/home.aspx); and
- Two North British built South African Railways giant 3ft 6in locomotives have been returned to Scotland –
  - No. 3007 a Class 15F 4-8-2 of 1945 at the Glasgow Riverside Museum of Transport (www.glasgowlife.org.uk/museums/riverside/Pages/default.aspx);
  - No.4112 a Class GMAM (Garratt articulated locomotive sub-contracted from Beyer Peacock of Manchester) 4-8-2+2-8-4 of 1956 at the Summerlee Museum of Scottish Industrial Life (http://www.northlanarkshire.gov.uk/index.aspx?articleid=15877).

4-8-2+2-8-4 No.4112 of 1956 at Summerlee Museum of Scottish Industrial Life.
Courtesy of CultureNI Museums & Heritage

The steam outline Alan Keef loco departs Alford on another journey along the Alford Valley Railway.

Iain McCall

'New' Operating Lines

Three new narrow gauge lines have been developed in recent years to both commemorate the industrial operations in Scotland and to provide a narrow gauge 'experience', and a further line is used as an attraction within a larger site.

The Alford Valley Railway is a volunteer-operated 1.35km 2ft gauge line which is in Aberdeenshire. An original line was built in 1979/80 using salvaged equipment from a nearby peat railway at New Pitsligo; but this was replaced by the present line in 1984 which runs from the former Alford Station to Haughton Park. A variety of locomotives have been used on the line since an original 1930s Lister diesel-mechanical came from New Pitsligo. At present there are three diesels available, of which one has a 'steam-outline'. One of the two available carriages incorporates part of an 1895 Aberdeen tram. Services operate on weekends from the end of March to the end of September, daily in July and August and for a couple of weekends in December. More information can be found at: http://www.alfordvalleyrailway.org. uk/content.asp?cpage=History. Close to the railway are two other attractions: the Grampian Transport Museum and the Alford Heritage Centre.

The 2ft gauge Leadhills & Wanlockhead Railway is a volunteer–operated line in South Lanarkshire that has been constructed since 1986 for about 1km from Leadhills to a terminal halt at Glengonnar along the trackbed of part of a former Caledonian Railway branch which was closed in 1930s. It is, at 1,498ft, the highest adhesion railway in Britain. There are seven ex-industrial diesel locomotives based at Leadhills

Hunslet diesel-hydraulic of 1975, No.6 CLYDE, at Leadhills on the 2ft gauge Leadhills & Wanlockhead Railway on 29th August 2005
A. M. Hurrell

from a variety of builders: MotorRail, Ruston & Hornsby, Hunslet, Hudswell Clarke and Clayton. There are also two steam locomotives in store: a 1913 0-4-0T from Orenstein & Koppel and a 1915 0-4-0T from Decauville. Four carriages are in use which are built on ex-peat wagon underframes and a former Simplex locomotive. Services operate at weekends from Easter to the end of September and for Santa Specials. Further information can be found at: http://www.leadhillsrailway.co.uk/ . Close to the railway can be found a Museum of Lead Mining.

The 2ft 6in gauge Almond Valley Railway operates for 0.5km from Livingston Mill to Almondhaugh and was developed between 1992 and 1998. It is part of the Almond Valley Heritage Centre which preserves and interprets the history and environment of West Lothian, particularly focussing on its rural past and the shale oil industry. Within the shale oil museum can be found the surviving 1902 electric locomotive from the Winchburgh Railway. The Almond Valley Railway has taken its inspiration from the allied histories of the shale oil / explosives / defence industries with its track and stock all drawn from these backgrounds. Three locomotives have been used to haul passenger trains: a 1941 'flameproof' Hunslet, a 1970 Andrew Barclay and a 1980 Baguley-Drewry. In addition on the site can be found two other diesels (Hunslet and MotorRail) and a number of battery-electrics (1941 Greenwood Batley and 1970s Brook Victor). These locomotives all came from Scottish explosives/defence sites or allied sites in England. Three of the passenger carriages used are built on former trucks from Bishopton, and one is a former vehicle from a navy munitions depot. The Heritage Centre is open daily except 25 & 26 December and 1 & 2 January. Railway passenger services operate at weekends from the beginning of April to late in October; daily in July, the first half of August and part of October; and at weekends in December. Further information on the railway, and downloadable comprehensive booklets on the Railway and its stock, can be found at: http://www.almondvalley. co.uk/Railway.html.

One further narrow gauge railway is located in the East Links Family Park near Dunbar, East Lothian. East Links is a 'farm themed activity park', which has 'a train safari' circuit of about 0.9km through the animal paddocks. The 2ft gauge railway was established in about 2000 using a 1938 steam-outline 40hp Baguley 0-6-0 petrol mechanical locomotive, previously used at Trentham Gardens (Staffordshire) and Alton Towers (Staffordshire), together with four bogie toastrack carriages from the same locations. The Park opens daily with railway operations. Information about the Park can be found at: www.eastlinks.co.uk/.

The 2ft gauge East Links Family Park Railway with its 1938 Baguley petrol loco and toastrack carriages (all previously at Trentham Gardens and Alton Towers)
Courtesy of East Links Family Park

General References

(there are articles on various individual lines such as that at Winchburgh in a variety of magazines, etc.):

Bridges A. (Ed), *Industrial Locomotives of Scotland – Handbook N*, Industrial Railway Society, 1976 (together with many other regional handbooks published by the Industrial Railway Society).
Cocroft W.D., *Dangerous Energy*, English Heritage, 2000.
Evans D., *Arming the Fleet*, Explosion! Museum / English Heritage, 2006.
Knox H., *The Scottish Shale Oil Industry & Mineral Railway Lines*, Lightmoor Press, 2013.
Macmillan N.S.C., *The Campbelton & Machrihanish Light Railway*, Plateway Press, 1993.
Nicolson M. & O'Neill M., *Glasgow – Locomotive Builder to the World*, Polygon Books / Third Eye Centre / Springburn Museum / Glasgow District Libraries Publications Board, 1987.
Robertson C.J.A., *The Origins of the Scottish Railway System 1722-1844*, John Donald Publishers, 1983.
Scotney D., *30inch Railways Worldwide*, Stenvalls, 2013.
Wear R., *Barclay 150*, Hunslet-Barclay Ltd., 1990.
Wright J. & Mackean I., *Circles Under the Clyde*, Capital Transport, 1997.

Acknowledgements

Thank you to the following for their help in the preparation of this article: Robin Chesters (Almond Valley Heritage Centre), Gordon Edgar, Karen Gallagher (North Lanarkshire Council), Patricia Grant (Mitchell Library), David Hall, Ray Hooley, Isobel Maclellan (Mitchell Library), Justin Parkes (Summerlee Museum of Scottish Industrial Life), Tara Paxton (East Links Family Park), Suzanne Rough ('Glasgow Life'), Hamish Stevenson, and Michelle Watt (SPT).

Thanks must also go to the Transport Research Institute at Edinburgh Napier University for allowing me to use their academic research facilities.

OVERSEAS INTERLUDE 1: DURANGO & SILVERTON
by Dave Collier

February 2015 was spent in search of steam in the snow... Not having been there before, my wife and I decided to go to Colorado to do the annual winter photography charter on the Durango & Silverton. Unfortunately, as can be seen, the area was in the middle of a freak "warm spell" with temperatures of +16C. Nonetheless, the charter was well organised, with many runpasts in the superb lighting providing outstanding opportunities for photography. Here 'K28' 2-8-2 No. 473 crosses the Animas River on the spectacular "Highline" section.

Author

Above: 473 running through the meadow at Shalona.

Jackie Fisher

Below: A little further on, and 473 traverses the rock face at Shalona Lake.

Author

Above: Another view of 473 running through the Animas River gorge.

Jackie Fisher

Below: The Rocky Mountains live up to their name as the charter passes MP 476.5 near Cascade Canyon.

Author

Above: Our final view of the photo charter shows it near Tacoma.

Author

Below: More normal fare on the D&S is illustrated by this view of a service train at Harmosa. Again the train loco is 473.

Author

A 15" GAUGE REVIVAL - A YEAR ON
by the Friends of the Thorpe LR

The story of the railway's renaissance in the first "Summer Special" had a heartening review in "Steam Railway": '...but standing out in the restoration category is the section on the Thorpe Light Railway, a 15 in gauge line that you may never have heard of, and yet has as interesting a story of revival as any other preserved line.' We had cleared fallen trees, dug out blocked drainage channels flooding the line and cut back vegetation to show the railway that was under there somewhere. We increased from one to three open days plus a society visit in both years. Not much work needed, then? Far from it, especially when dealing with 45 year old infrastructure .

Revival of a railway, even a 770 yard long 15" gauge one, relies on people for maintenance, development and operation. At the start of 2014 we had around seven regular volunteers plus a few more who helped at times; by early 2015 we had about a dozen. On our three open days a number of new Friends joined. Some are local and remember the line from its Whorlton Lido Railway days while others are railway enthusiasts. As well as current and former NRM York and Locomotion Shildon employees volunteering in their spare time we have a number of Locomotion volunteers too. Our volunteers now span the age range from teens to sixties. As well as their efforts they bring new skills plus useful tools and equipment on the regular weekend working party days. Lighter summer evenings allowed useful midweek work to be done.

After over four decades some of the infrastructure needed attention. First was completing the platform lowering-not a fault but needed because the current Severn Lamb carriages are much lower than the original Whorlton ones. Half a dozen flagstones at the platform's east end were cracked so new ones were bought and delivered to the shed. These were lifted onto one of the bogies from a dilapitated Whorlton coach, hand-propelled to the station and put in place. Second is the replacing the rotting wooden sleepers under the two station points (most sleepers are metal.) Of concern was the point next to the platform so a speed restriction was put on it. After consultation with several other lines recycled plastic sleepers were used. These should survive better than wood in the damp environment. November's working party removed the old sleepers, drilled the plastic ones using loaned generators then put the plastic sleepers in place. Work on the other point began in December. A third issue is to restore the correct level of ballast in places. In two locations on the near loop it has dropped. So in December measurements were taken of the cant around the loop to restore it early in 2015 to the super-elevation created by the men who laid the track in 1970.

While we had cleared much by opening day in June 2013 there were still eyesores from previous times. Where the near loop begins its curve in the south west corner

A driving training session with SOONY on August 23rd as it circuits the near loop. Note that the overgrown grass in the middle of the track has all been cleared.

All photos by Philip Champion

near the road had become a bit of a jungle. At the impetus of the original owner's grandson this was cleared in September. It revealed the collapsed remains of the original wooden engine shed moved up the bank in the 1990s by the line's second owners when a new shed was built. Set on breeze blocks it had become a stable for several horses. As the 'jungle' was cleared more of the long, tall retaining wall was unveiled with young branches which had started growing in it now removed. The wall probably dates back to the c1830 turnpike road. It is now more of a feature for passengers to see.

Another eyesore went in October: the former c1990s pheasant pens where the near loop tracks diverge. This had been a gloomy area of netting, overgrown branches and drooping bits of wood in between the tall fir trees. With a good number of volunteers this was generally cleared in a day, the soil raked over and the whole area made more pleasant.

A major restoration still to come is the tunnel roof. The original, built of sleepers, was removed in 2012 as several had dropped down at alarming angles. A grant received in November is paying for a new concrete one designed by one of our newer Friends, a retired architect who lives in the nearby Whorlton village. January 2015 saw delivery and movement across site of the various concrete parts on three days. Preparatory work included removing the crumbling arch at the far end leaving just the frame, made of curved rails. Photos were taken to help with rebuilding this.

While some working parties just require pairs of hands, tools and maybe a wheelbarrow others have benefitted from BESSIE. Effectively this is a motorised

small dumper truck assembled in 2004. Its skip has proved valuable in track work, transporting cut branches plus the bits of metal, uncovered during September's 'jungle' clearance, which made it easier to load them into a van for waste disposal and recycling. BESSIE has seen used both at weekends and on the summer evening working sessions. It can be seen in the background of the photo, left. It proved too to be a good passenger locomotive for our June and July open days. Though able to pull more it hauled the open Severn Lamb carriage as trains were according to demand rather than a fixed timetable. It also was used for an October visit by a railway society.

It was decided to widen the operating experience of the volunteers at each open day. While the Friends' Secretary - the person who really saved this line - was usually Responsible Officer work, holidays meant that a couple of others with railway operating experience have stood in. As all volunteers sign for their copy of the Operating Procedures on open days they rotate around different tasks, with operating ones supervised initially by those with railway experience. So a volunteer could be front of house, selling tickets or Guard. Not only does this give variety but helps motivation - much better than say banishing someone to the entrance gate by the road for the five hours of the open day.

The highlight of the operating season was the return of steam. For our reopening day in June 2013 SMOKEY JOE came from Sherwood Forest. Before the 2014 August Bank Holiday a Kirklees Light Railway team in Yorkshire drove to the Perrygrove Railway in Gloucestershire to pick up new build Baldwin-type SOONY. Unloading at Whorlton took eight people a good hour or two. When it was steamed on the Saturday morning the sight and smell of steam was a joy to behold. It undertook eight driver familiarisation circuits with the two roofed Severn Lamb carriages. SOONY proved to be very responsive with plenty of power, a delight to drive. What was surprising was the gradients you were not aware of but which became apparent when driving the steam locomotive. Sunday's operations were enhanced by a barbecue put on by the site owners' for the local people and railway passengers. At first trains alternated between BESSIE with its open carriage and SOONY with its two carriages but after lunch all trains were steam- hauled running to demand which was constant most of the time. By the end of the day SOONY had done 27 circuits and BESSIE nine. After the last train departed SOONY did a final run for the volunteers and was loaded up for transport to Kirklees.

During its visit to Whorlton SOONY did 35 circuits to run 15.4 miles. it is estimated that BESSIE had done the equivalent of 74 circuits in 2014 which is 32.6 miles.

It was clear that track levels needed attention on the straight and some places in the near loop. The line is super-elevated in most of the latter but has dropped in places and at one point the cant went the other way. A track improvement programme started in January 2015. Two lengths worst affected were dug out completely from mud and old ballast, the track jacked to restore the correct cant and transitional curve then a fresh bed of ballast laid and packed. BESSIE transported the ballast to the work site then moved slowly back and forth over the reballasted section a number of times to help it settle. To prevent this area being waterlogged again a 40 foot long drainage trench was dug in the embankment behind to divert water running off the land the other way into the main drainage channel feeding the lake.

Another driver training session with SOONY on August 23rd, as it goes past the station on the 'main line.'

Unexpected problems need to be dealt with. Late last year a volunteer noticed the near loop curve leading into the station had more cant than usual. This February we dug out part of the curve to find a tree root going under one sleeper and lifting the track (*right*). It was 6" long as it went under the sleeper, becoming 10" wide and at least 6" deep with half a dozen smaller roots. Much sawing was done but the main root could not to shifted until a chain saw was used another day. The original owners never thought of this when they planted saplings 45 years ago! Now the curve can be restored to the correct cant.

A very full year then for a small line with plenty of work to come. We have been offered the loan of another i/c locomotives plus one of two steam locomotives for a visit. It will be interesting to see what 2015 brings to the this 15" gauge line in the south County Durham countryside.

Checking the cant of the reballasted section in the near loop using an old Whorlton Lido Railway coach bogie in January 2015.

GARRAWAY ENCOUNTERS
by Vic Mitchell

'AG' At Harbour Station in the 1960s.
FR

I became fascinated by the Festiniog Railway when studying abandoned cigarette cards as a young child. In later years when I read of a meeting to discuss its reopening, I hurried to Bristol in September 1951. It had been called by Leonard Heath-Humphries, another teenager, and it was here that I first heard AG speak and it was with such authority. A committee was formed and they were both on it.

Later, I was invited to attend with a view to joining it, but Garraway challenged my suitability. My earlier promotional efforts with H-H won the day. (In those days Christian names were limited to family use.)

He soon left to work on the FR full time, while I became a student at Guys Hospital for six years. This was ideal, as all FRS meetings were held in London and I became a director when the company was formed in 1954.

However, the FR was too far away and Garraway encounters ceased. When my military service deferment ended, I requested an appointment in Wales and was given one! Barbara and I spent one weekend per month on the line, she enjoying helping on the buffet car.

I explained to Allan that I felt that directors should have experience of working in all departments. Thus encounters were revived, as he was often guard and I assisted him. More often than not he was the driver and I fired to him. My tasks ranged from the booking office to Boston Lodge, all under his supervision in those days. All were fascinating and fun - FF, later to become JGF - Jolly Good Fun.

I developed a clear acrylic casting business called Mitchell Mouldings and he obtained small cast metal FR loco models for producing giftware. This was an unexpected encounter.

Lack of cap badges presented a problem and I suggested that we could use those of the Royal Engineers. He was shocked until he grasped that R and E only needed reversing and the foot of the E cutting off. (I still wear one).

Pleasant encounters were had in Sussex, when he stayed with us and gave talks to our local FRS group, HANSAG. Reminiscences abounded and valve setting on my locomotive was perfected.

There followed an annual visit by Barbara and I to stay with Allan and Moira, after they had moved to Scotland. He had built a three floor spacious bungalow overlooking the Strathspey Railway. The ground floor was for storage, the middle one for dwelling and the top was paradise - all 0-gauge, built by him since teenage. You were given a station to manage, with full block signalling and bell codes. Sadly I received many aggressive encounters, due to incompetence.

Also on this floor were cases of films and photographs, dating back to the 1930s. I saw many of the former, but he

On the footplate of LINDA in June 1966.
Gil Roscoe / FR

would not allow them to be copied. Barbara and I had set up Middleton Press in 1980 and he asked if we would publish his Garraway Father and Son. This we did, using many of his pictures and later even more appeared in our albums.

Thus our annual encounters contained many pleasures of a unique nature, also others from red squirrels to a steam outline windvane all outside his kitchen window. Of lasting enjoyment for all are the albums to which he contributed so extensively - Branch Lines around Portmadoc 1923-46, Branch Lines around Porthmadog 1954-94, Porthmadog to Blaenau 1955-95, Festiniog in the Fifties, Festiniog in the Sixties, Festiniog - 50 years of Enterprise, Festiniog 1946-55 - The Pioneers' Story, Ffestiniog in Colour 1955-82 and Return to Blaenau 1970-82. He only appeared on the cover of my DVD, Festiniog Revelations. However his memory will live for ever.

PRINCE carries a wreath on December 30th 2014 to mark the death of Allan Garraway.

Stephen Greig / FR

2014 GALA MEMORIES

We begin this year's selection of Gala images in Devon, at the Devon Railway Centre, where Barclay JACK joined regular visitor PETER PAN for the August event.

Alan Frewin

Remaining in Devon, we travel north to the Lynton & Barnstaple, and see visiting Bagnall 4-4-0T CHARLES WYTOCK and resident 0-4-2T ISAAC lining up at Woody Bay during the Autumn Gala on September 27th.

Andrew Budd

Moving to Eastern England, first we visit Page's Park station in Leighton Buzzard, where newbuild Kerr Stuart 'Wren' JENNIE waits with classmate PETER PAN.

Peter Donovan

Further southeast, and GREEN GODDESS reverses into a siding at New Romney, whilst WINSTON CHURCHILL awaits the 'right away' with a service for Hythe on September 27th.

Alan Frewin

The Twyford Waterworks Railway Gala in June each year provides ample opportunities for industrial railway photography in this rural setting. The loco at the head of this rake is resident AYALA.

Iain McC

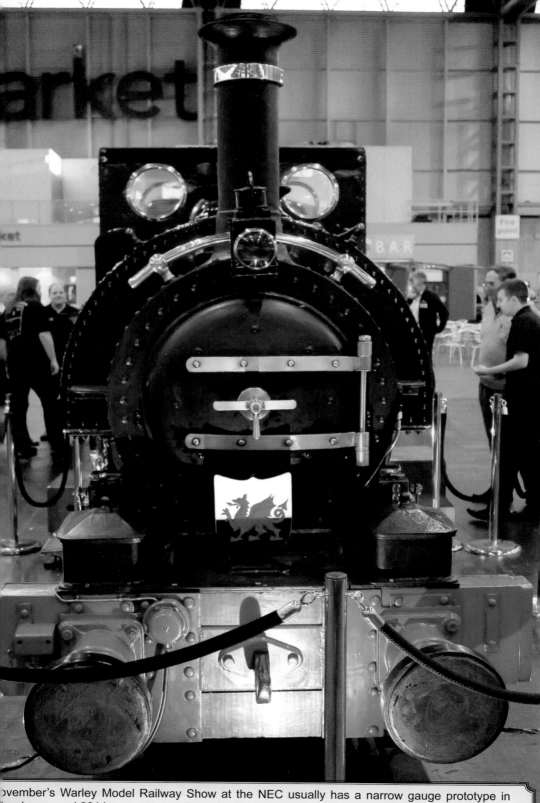

November's Warley Model Railway Show at the NEC usually has a narrow gauge prototype in attendance, and 2014 was no exception, with TALYLLYN promoting the TR's 150th celebrations.

Keith Frewin

SRAGI 14 (O&K 10750/1913) creates a powerful image at Statfold on 29th March.

0-4-0VBT LEARY shunts slate wagons during the Llanberis Lake Gala on 5th July.

Both: Peter Donovan

Us narrow gauge enthusiasts are almost conditioned to headlines about ex-Penrhyn Hunslets, but it was ex-Penrhyn Ruston R J BROWN that made the news on August 2nd, when it returned to service during the Tinker's Park Rail Gala.

Alan Frewin

Our final view from this selection is of Alan Keef's September Open Day, and WHHR based Motor Rail 264 of 1916. Staring at the camera from the leading bay of seats in the coach is regular *Narrow Gauge Net* contributor Andrew Budd.

Both: Peter Donovan

BIG PLANS FOR THE LITTLE MINE

by Philip Conway-Jones

The run round loop on the LBLR, seen in 2014. The loco on the left is resident Motor Rail 21282/60, which after a somewhat peripatetic existence arrived on site in 1993, whilst on the right is Hunslet 7446/75, originally delivered to a rather different mine, Prince of Wales Colliery in Pontefract.

All photographs by the author unless otherwise stated

Winning gold was the original aim of mining at Lea Bailey in the Forest of Dean. Nowdays, the focus is winning funding and supporters, and countering resistance over a project to extend the two foot gauge railway and preserve heritage mining and narrow gauge equipment.

The Bailey Level opened in 1906 under Wigpool Common, NGR SO 6480 1980. The Chastan Syndicate was floated and raised £75,000 in £1 shares, incredibly based on finds of 6 grains of gold per ton. Nothing came of it, and the mine was closed one year later.

It reopened in 1920 for the extraction of 3,000 tons of iron ore, and the level was eventually driven 580 yards into the hill, with two headings of 19 and 44 yards.

The mine was abandoned for many years until the gale was aquired in the 1990s by Ray and Jonathan Wright, owners of the highly successful Clearwell Caves visitor attraction. They were also responsible for Wigpool, Old Ham and Old Bow iron mines. (A gale forms the right to open a mine within the hundred of St. Briavels).

The Wright's plan was to open a Lea Bailey Gold Mining Experience, complete with narrow gauge railway, but the project was put on hold due to a number of setbacks at the site, which due to its remoteness, suffered from a spate of theft and vandalism.

Nature began to reclaim the site, just as at many other Forest of Dean coal and iron mines, where keen explorers can still spot relics in the undergrowth. It was the discovery of such interesting artefacts protruding through the bracken that led several Royal Forest of Dean Caving Club members, including Rob Needham, to approach the Wrights in 2012 and ask if they could restore the site with volunteers.

Lea Bailey consisted of the mine level, an old shed constructed from corrugated iron and panels from a water tank, and approximately 50 yards of overgrown 30lb two foot gauge rail with a siding into the shed. Beyond this, along the bed of the former Mitcheldean Road & Forest of Dean Junction Railway was roughly laid a further couple of hundred yards of heavier weight rail.

The Lea Bailey Light Railway Society was formed in 2013 with a mission "to install and operate, in stages, a two foot gauge railway between the former Lea Bailey gold mine and Drybrook on the trackbed of the former Mitcheldean Road and Forest Junction Railway." But restrictions were imposed on the size of operations by landowners the Forestry Commission, meaning that a "Stop" had to be placed in the rail after around 40 yards, leaving a tantalising 200 plus yards untouchable and an operating area purely within the mine site.

It is hoped that the Society can garner support from the Forestry Commission and Deputy Gaveller (who administers the gales) to overcome this obstacle and head towards Drybrook and Hawthorns Tunnel.

Not that there wasn't already enough to do on site. The undergrowth had to be cutback to expose the rails and sleepers and the shed had to be made secure and weatherproof. The first loco to arrive at Lea Bailey was a Simplex 4wD MR21282/60, loaned by Clearwell Caves. This was the start of much kindness and support shown to the society by Ray and Jonathan Wright from Clearwell.

An old shipping container was purchased to store the Simplex, which became a valuable workhorse shunting an ever-growing collection of ex-NCB rolling stock around the site. At the same time work was going on behind the scenes to restore an underground Hunslet 4wDHF 7446/75 to working condition. This was stored securely inside the mine and also became a feature of regular weekend volunteer working parties.

Twice-yearly public open days became established, under strict operating rules to satisfy health and safety requirements. This meant that the society was operating according to a rule book and visitors were not permitted to ride on the locos or wagons. The evolution of the society and regulations

One of the interesting challenges of restoring a mines railway - tracklaying in a tunnel! This view is in Hawthorn Tunnel in August 2014.

have meant that for the time being, it is not a publicly accessible, passenger carrying railway, but more of a mine railway preserving historic equipment. This could change in the future, once obstacles have been overcome and we have more support.

However the society's activities have not gone un-noticed by the media and we have enjoyed great radio, TV and newspaper coverage. A useful public relations exercise involved taking the Simplex, U-tipper wagons and portable rail to the Welland Steam Fair in 2013, where a face shovel and traction engine-powered rock crusher were also seen in action.

It soon became clear that a loop was required at Lea Bailey, as the track layout did not allow a loco to run round a wagon. Two sets of Y points and sleepers were moved from storage in Hawthorns tunnel to build this in time for another very successful open day. Towards the end of 2013 plans were taking shape to restore a Wingrove & Rogers WR5 works no L1009 and an Eimco 12B rocker shovel, both laid up at Clearwell.

An appeal was launched for donations for 100 tanalised sleepers to replace rotten ones which were being exposed as the track was being dug out. Thanks to donations and society funds, 150 weighing around three tons, were purchased. Much of the trackwork was overhauled in time for the summer of 2014, which saw a narrow gauge feast in the Forest, with special events at Lea Bailey, Clearwell, and Perrygrove railways over a weekend. The battery loco was overhauled in time for a demonstration at Clearwell in which society members helped haul a wagon from the depths of the mine up to the surface, where it was towed out to daylight.

New on the scene for restoration at Lea Bailey was a four wheel Tredomen man rider, made at the former Powell Duffryn workshops in Caerphilly. There are a further four at Clearwell from different manufacturers. A further four locos also joined the expanding collection - two 1977-built WR18s: WR7888 of and WR7964 of ex-Frazers Hush fluorspar mine in Weardale, a Clayton 13/4 ton 4wBE (works no 5961C) and a ¾ ton Logan 4wBE no 1066. All require various levels of restoration.

And then on one day in 2014 we were given a reality check when the Mines Inspector came to visit. As a result of his report, all locos were moved from storage in the mine and the Hunslet had to be returned to Clearwell. No locos were allowed inside the mine, and rolling stock had to be moved inside with a long coupling bar. The arrival of 10 tons of limestone was just the start of a long programme of ballasting all the track.

The Land Rover assisted process of moving one of the latest arrivals, WR7888R, from its unloading point up to the rails.

THE LEA BAILEY TEAM - THEN & NOW

Above: A postcard of the gold mine and its workers in former days.

Unknown, Author's Collection

Below: The current generation of Lea Bailey workers is seen in this group photo celebrating the movement of WR7888R (see photo on opposite page), in January 2015.

Author

Ballasting in progress on the "Long Siding", in March 2015

But enthusiasm wasn't dampened as another exciting development was the start of track laying in the 638 yard Hawthorns tunnel, further along the line from Lea Bailey. This is also owned by the Wrights, who laid around 200 yards of two foot gauge track in the 1990s. The plan was to get the two 4wBE locos and the Hunslet working a single track with sidings inside the tunnel. Society members have so far laid a further 50 yards with enough track inside the tunnel to complete another 100 yards. However, access issues have currently put this on hold.

The collection grew bigger still in the spring of 2015 with the acquisition of two more battery locos. Murphy Construction was selling some of its fleet and the society jumped at the chance to acquire a rare two-foot gauge WR8 and also a an 18" WR5 which can be used as spares for the other WR5 or restored to visit other railways as it is light enough to sit on a trailer behind a Land Rover.

Looking ahead this summer, the LBLRS is hosting part of the Narrow Gauge Railway Society AGM on May 9th, with a public open day on May 10th, where it is hoped to have the Eimco rocker shovel in action, the Simplex, at least one battery loco and Roy Etherington's 0-4-0CA ISSING SYD from Statfold.

The challenges ahead include the purchase of more sleepers for Lea Bailey, the completion of a long siding behind the shed, and further work at Hawthorns tunnel. Storage solutions such as shipping containers will have to be found for the new locos and rolling stock. Members have also offered to help with the restoration of a small area of track at Hopewell Colliery Museum, which is being re-opened complete with Ruston loco.

"Over the next year or so the society will face some serious challenges," said Chairman Rob Needham. "We are in need of secure weather and vandal-proof accommodation at Lea Bailey for locos and tools."

"To rebuild the tin shed is going to cost us at least a couple of thousand pounds. Track-laying in the tunnel will run out of rail at least 150 yards short of the far end, so we'll need to buy rail – not cheap, especially with delivery costs. We have almost used all of the 150 sleepers. A similar quantity will cost over a thousand pounds.

"We are also near to running out of rail at Lea Bailey, so will need to buy more rail (there is no point trying to use the heavy rail at Lea Bailey until we have suitable lifting gear and the resources to hire a hydraulic rail-bender).

"When we get the WR18 battery loco ready to run we will need to buy six 12v batteries, which will cost over £500. Then we'll need solar panels on the shed roof to charge loco batteries (4 are OK to take home, but another 6 is getting a bit too much.) And there will be more costs – when we have the shed completed we will need to equip it.

"So it looks likely that over the next year we will need something like five thousand pounds, and that excludes the normal day-to-day running costs such as diesel fuel, fishplates, nuts and bolts, ballast for the track, etc.

"We also need to make friends with local councillors and MPs and anyone else who can help us. I think that we need friends who will support us when we talk to the FC about using or leasing the trackbed from Lea Bailey towards where Mitcheldean Road or Drybrook."

To join the society and keep in touch with our activities, or make a donation, please go to: www.leabaileylightrailway.co.uk or like us on Facebook at: www.facebook.com/leabaileylightrailway

What the small band of volunteers has already achieved at this site can be clearly understood when comparing the view above, dating from July 2012, with the other views in this article, many of which include the shed, an original feature of the site from its industrial days.

DOLGOCHS' COATS OF MANY COLOURS
by Iain McCall

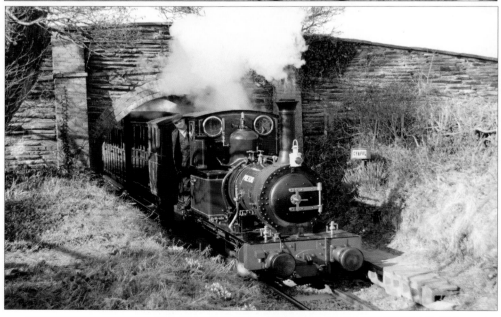

Talyllyn Railway No. 2 DOLGOCH has appeared in several colour schemes since its Steam Railway sponsored restoration in 2011. It is seen (*top left, Iain McCall*) on its return to service special in May 2011. Whilst in this livery, it was temporarily renamed PRETORIA for a photo charter, pictured (*below left, courtesy TR*) at Cynfal. The TR's Christmas Surprise for us all was the reliverying of Nos 1 & 2 into the original TR livery (*top, Darren Turner*). During the same period, the TR Garden Railway's DOLGOCH has also appeared in "standard green" at Llechfan (*centre right, Iain McCall*), and "early preservation green" at the Llangollen's Gala (*bottom right, Phil Budd*),

GLOUCESTERSHIRE TO WEMBLEY & RETURN!
by Stephen Mourton

Introduction

There are very few genuine narrow gauge coaches extant in England. Many narrow gauge railway coaches were built by various firms, but mostly for export and they served on railways in countries all over the world. In England, there were a relatively small number of passenger carrying narrow gauge lines and the few carriages that survive in preservation have been heavily rebuilt with hardly any of the original material remaining.

So there was a sense of excitement in 2007 when we heard that a friend of a North Gloucestershire Railway Director had by sheer chance discovered what appeared to be a narrow gauge coach in the garden of an unoccupied tin bungalow between Aston Cross and Teddington Hands, Tewkesbury. We hastened to visit the site, just over six miles from Toddington, and it certainly looked like a coach hidden behind the bungalow, albeit heavily overgrown and full of junk, with parts rotting away. It was not easy to get close enough for further investigation. Contact was established with the site owners, Overbury Estates, who acknowledged our interest, but were not ready to proceed further at the time.

Then early in 2011, with the land about to be sold for redevelopment, Overbury made contact with the NGR and the coach body was offered to us as a restoration project. However the previous occupiers had covered the roof with corrugated iron sheets, which kept the worst of the weather off the body with the roof remaining intact and in excellent shape. In addition the floor was raised off the dirt on a two-wheel trailer which had been used to transport the coach to the site, with the result that air had been able to circulate underneath, stopping the rot taking over. There was even an intact pane of glass in one of the windows. More remarkably, one end of the coach interior still had a worksplate, revealing its builder to be Gloucester Railway Carriage and Wagon Company (GRCW). This was a huge help in researching the history of the vehicle, as we were pretty clueless about the coach and where it had originated, speculating it might have come from a tramway rather than a railway. Also with no rail wheels, the gauge on which it had run was a matter for more guesswork.

Historical Background

Much of the GRCW archive is held at the County Records Office, Gloucester and subsequent research in the Board Minutes revealed the coach was one of eighteen constructed and delivered in April 1924 for use at the British Empire Exhibition held at Wembley in 1924 and 1925. Fortunately the archive also included a works photograph of one of the coach bodies, which were built to the order of 'The Stronach-Dutton Road Rail Company'. This Company won the right to operate a two foot gauge

AT THE WEMBLEY EXHIBITION....

Above: A rake of the Gloucester coaches being hauled by a Halley road-rail tractor in 1924.

Below: A glimpse of another Gloucester vehicle waiting at the platform behind a Beardmore Sentinel loco, also in 1924.

All photographs North Glooucestershire Railway Collection

railway at Wembley to its own unique patent - not to be confused with the more well-known 'Never-Stop Railway' which also ran at the Wembley Exhibition. GRCW built the bodies and mounted them on ex-War Department Light Railway class D wagon bogies and frames; GRCW constructed many of the latter, so presumably had lots of surplus bogies lying around in the early 1920s.

The coach measurements quoted by GRCW are:

Length over pillars 17 ft 8.5 inches;
Width over corner pillars 5 ft;
Length over buffers 20 ft 6.5 inches;
Width over step boards 6 ft 3 inches;
Centre of bogies 13 ft 9 inches;
Height from rail to top of roof 9 ft;
Seats 24 passengers;
Weight 1 ton 13 cwt.

Tens of thousands of people visited the British Empire Exhibition and many would have ridden in our coach. The Roadrails line was double track and 1.25 miles long; it operated for 21 weeks in 1924 and carried 300,000 passengers. There are pictures of Roadrails trains in action which show the coaches in sets of three or four. The trains were hauled by Halley petrol tractors either with pneumatic tyres on the rear wheels running on paved strips outside the rails or with caterpillar tracks. The front axle of the tractor was mounted on a bogie running on the two foot gauge rails. There were also two Beardmore-Sentinel steam powered tractors which hauled the coaches.

After the Exhibition most of the material at Wembley including the 'Never-Stop' railway was auctioned off, but apparently not the coaches used on the Roadrails system which the archives indicate were always owned by GRCW. Roadrails Ltd went into voluntary liquidation in 1927. We do not know what happened to the coaches after the Exhibition, except for the one we now own. Overbury Estate was in contact with a lady who was related to the last occupier of the bungalow and land where the coach resided. We learnt from her that relatives acquired the coach body in the early 1930s and that by 1933 it was installed on their farm at Roberts End in Gloucestershire where the family used it as a summer residence when they let their farmhouse to earn extra income. Then before 1939 it was moved again by one of the family to the site at Teddington Fields where it remained until removed to Toddington. It was still used for family visitors from time to time, but gradually turned into an outhouse.

Restoration timeline

February 2011

Work began on the rescue from the undergrowth where it had lain for over seventy years. This proved to be a challenge in itself as brambles, trees and other vegetation were very much in evidence. However the volunteers managed eventually to free it from its shackles.

March 2011

The coach was craned onto a HIAB lorry and transported to Toddington, where it was transferred to a pair of flat trucks. The coach was at last back on the rails.

WHAT TWO YEARS OF HARD WORK CAN ACHIEVE....

Left: The coach pictured shortly after arrival at Toddington in March 2011. Only the corrugated iron roof and temporary doors have been removed.

Right: The opposite side of the coach viewed two years later in March 2013. The extent of the external restoration carried out by the small group of volunteers can be readily appreciated.

There followed a short journey to our depot where the boarded-up doorways were uncovered to reveal a basically sound body frame. The corrugated iron roof sheets were removed and it was confirmed that the roof had survived in excellent condition. The outer beams of the frame and the floorboards proved to be beyond repair and were consigned to the wood scrapheap. The main part of the frame, however, was in excellent condition. No further progress was made during the 2011 running season.

September 2011

At the beginning of September, it was decided to split the body from the frame. First the body was braced with timber, bolted in place using the original grab handle holes in the doorposts. Then with much levering and jacking, the task was accomplished. Work then began on the frame. Rotten wood was carefully cut away and timber was ordered for replacement of the two 18ft side beams and the oak buffer beams. Some metal parts including numerous nuts and bolts were needed. A mortising machine was purchased to prepare the new timber. Over 40 mortises were cut with much careful measuring, (measure twice – cut once), because one mistake would put us back to step one. Incidentally it was observed that the holes and spacing of these on the frame confirmed that it had indeed once been on a World War One class D wagon.

December 2011

Around this time the volunteer team had virtually completed the underframe of the coach together with the cutting of all mortises and slots necessary to accept the coach frame and posts etc. All new timber was coated with wood preservative and anti-woodworm treatments.

January to March 2012

Thoughts turned to the coach floor and having received a favourable quotation from a local timber supplier the wood was purchased. Work commenced on laying the flooring which was incidentally of a tongue and groove nature. The team got to work and over the coming Sundays the floor was completed and slots cut to marry up with those cut in the underframe. The team then prepared to marry up the chassis and coach body which, after much toil and sweat, was finally achieved with the help of other members of the NGR crew.

November 2012 to March 2013

The body panels were stripped off and replaced. There was only one area of the body framework which needed attention. This small section was carefully cut out and an oak patch was spliced in. The whole outside of the body was then sanded, primed and undercoated.

October 2013 to March 2014

Having got the external bodywork in reasonable shape our thoughts turned to panelling out the interior where necessary. This was a fairly time consuming job as accuracy was necessary to ensure each piece of tongue and groove wood fitted tightly.

Having completed this somewhat meticulous job the next item on the agenda was to investigate the seating arrangements and supports.

Luckily the works photograph of the coach showed a bit of the seating and the shape of the supports. The location of the mortise joints established the exact position of the seats.

The seats were of the slatted kind and the backrests were the knifeboard type. The slats were fairly easy to put in and equally spaced to effect a tidy arrangement.

We tried very hard to copy the shape of the supporting pieces of timber for the seats. Detailed measurements of the seating width and the height from the floor were transferred to some tracing paper at full size. Using some geometric items to draw the curved portions, we at last had a shape that resembled the item in the picture. We put the tracing paper of the outline onto the wood and, hey presto, the job was a good one. We were well pleased with that.

October 2014 to March 2015

Much refreshed from another enjoyable running season we now had to turn our thoughts once again to the next phase of the Coach project.

It was decided to give attention to the windows of which there are ten. As mentioned earlier when the coach arrived on site one original piece of glass was intact - truly amazing given the state of the coach when found. All the window frames had the remains of very old putty which proved somewhat difficult to remove even with the aid of modern tools but this was overcome eventually. Once all the frames were cleaned, we ordered safety glass to BSI specifications and this was duly installed.

Attention was now focused on yet more painting given the fact that the outside and inside was now at the undercoat stage. We needed to ascertain what colour the coach was in its heyday. The big problem was the picture we had was a sepia print which basically showed dark colour only. Despite much research we were still none the wiser. So taking the bull by the horns so to speak, we chose Highland Green for the outside, with internal colour of Cookie Dough. The slatted seating was painted black and the knifeboard backrests in Rich Chestnut topped with a black strip of timber. The floor will have slats between seat rows and the whole lot creosoted for good measure.

The coach was originally equipped with lighting as two roundels were evident when the coach was found. The third one was missing so eventually one was made to compensate for the loss. Wiring was reinstated and three new lamp fittings were acquired and fitted. The wiring was routed back to a 12 volt burglar alarm type battery which is housed in a neat box fitted at high level within the coach. The power source would originally have been an accumulator. The covers for each light were of the bowl type and when lit give a very satisfactory effect.

The roof's canvas covering has been stripped back to await replacement and after that has been done a final top coat of paint will be applied to the exterior.

Other details such as grab handles for the coach doorways and full length running boards will need to be manufactured. Then chassis support bars must be made to fit along both sides. This work is scheduled for October 2015 as we now have to start the Running Season in order to generate yet more funds.

The Future

The Narrow Gauge Railway is currently sounding out methods of fundraising to purchase two bogies for the coach.

Anyone wishing to make a donation however small towards the purchase of materials required to complete this worthwhile project would be most welcome. Viewing the coach is possible by prior arrangement or on any running day, so helping to make our dream come true for this coach to 'ride the rails' once again.

The current state of restoration is apparent in this March 2015 view.

OVERSEAS INTERLUDE 2: STARS OF SANDSTONE

April each year sees a mecca for narrow gauge enthusiasts around the world, in the form of the "Stars of Sandstone" event, which takes place on the Sandstone Estate in the shadows of South Africa's famous Drakensberg mountains. The show features narrow gauge locomotives both large and small hauling a variety of trains, surrounded by spectacular scenery and other examples of South Africa's transport heritage. In these views we see (*above and top right*) 'Feldbahn' 498, and (*lower right*) Cockerill-built NGG16 No. 88 in Alfred County Railway green livery.

David Benn via SHT (2), SHT

RHEIDOL RENAISSANCE

by Will Smith

The Rheidol's Great Western renaissance is exemplified by this view of No. 8, in Great Western livery, and matching coaches, on 26th July 2011.

All photographs courtesy VoR

The ancient Kingdom of Wales is a place of myth and legend and a bit of a Mecca for narrow gauge enthusiasts for in this land there are at least a dozen narrow gauge steam railways, each very unique and with its own character.

The Vale of Rheidol is a steeply graded and highly scenic narrow gauge tourist railway running from Aberystwyth on the Welsh Coast to Devil's Bridge in the Cambrian Mountains. Laid to a gauge of 1' $11\frac{3}{4}$" it was originally built by a private company to carry lead ore from mines, it also was quick to establish itself as a tourist railway taking Edwardian holidaymakers to the beauty spot of Devil's Bridge, the area having been a tourist destination for hundreds of years with visitors flocking to view the spectacular Mynach Waterfalls.

After several years as an Independent company it wasn't long before it became absorbed into the Cambrian Railways in 1913. During the Great War, the line found a new purpose hauling army territorial troops between training camps which had been set up in the Rheidol Valley. The demand for motive power put pressure on the existing fleet so the railway hired PALMERSTON, an 0-4-0 tender saddle tank from the Ffestiniog Railway to ease the pressure on the existing locomotives. In 2014 the railway celebrated this historic occasion by hiring the locomotive once again. In September 2014 PALMERSTON worked its first passenger services on the VoR for over 90 years, a momentous occasion.

PALMERSTON is pictured on 23rd July 2014, on one of its test runs prior to the September celebrations.

The locomotive was found to be a very capable and reliable performer on the steep gradients and handled a short set of three carriages and brake van with ease. A temporary air braking system was fitted to ensure the locomotive was compatible with the Vale of Rheidol stock.

In the "grouping" of 1923 the myriad of smaller railway companies in the British Isles were amalgamated into just four large companies, the Cambrian became part of the Great Western Railway (GWR) whose empire stretched all the way from London Paddington. The Great Western were quick to exploit the tourism potential of the line and scrapped the railway's original Davies and Metcalfe stock building three brand new locomotives at Swindon Works. Authorisation was only given to build two locomotives, so some clever accounting was necessary to ensure the third was put through the books as a re-build of one of the originals!

It was during the 1920s that freight traffic ceased and the GWR took the bold decision to operate the line purely as a tourist attraction - the line flourished. The railway's current owners regard this period as its heyday so locomotives and stock are presented to give a flavour of this era.

1948 brought the nationalisation of British railway network, but the railway ploughed on and continued to operate on a seasonal basis. Despite closure threats during the 1960s, the railway soldiered on and through the dedication of the management it survived the infamous "Beeching Axe" at a time when the British railway network was being rationalised and steam phased out.

The line gained some notoriety by being British Rail's only steam operation. In the 1970s, problems with lineside fires, caused by badly maintained locomotives and poor quality coal, prompted a move to oil firing. British Rail ownership was to last until 1988 The line's future was discussed in Parliament as it was felt to be inappropriate that British Rail should still be operating a purely tourist the line. The line was put up for sale with the current company being the successful bidder.

Now owned by a Charitable Trust, the Phyllis Rampton Narrow Gauge Railway Trust, the railway has had to work hard to overcome the many years of underinvestment. Upon taking over from British Rail, it was soon realised the three locomotives were in a poor state of repair, the carriages were desperately in need of overhaul and the condition of the track left much to be desired.

The new company launched a project to re-lay the entire twelve mile track, using heavy rail to ensure that it is fit for purpose for many years to come. Two of the three locomotives, No 8, LLYWELYN and No 9 PRINCE OF WALES received heavy overhauls at the Pant Works of the Brecon Mountain Railway. These two were converted back to traditional coal burners in 2012/13 in the works at Aberystwyth. Unusually for a tourist railway, the Vale of Rheidol is operated by a small team of staff. Most staff are multi-skilled so one day may be engaged in Permanent Way duties, the next out Driving, Firing or Guarding.

Over recent years as well as running the regular services, the railway has launched ambitious programmes to restore the line. This started with the restoration of intermediate stations along the route to something more akin to their appearance in the 1920s and 1930s. Replica station buildings reminiscent of the originals dating back to the railway's opening were constructed and add to the character of the stations. The original buildings had fallen into a state of disrepair and were removed during British Rail's ownership of the line. Another distinctive feature of the line was the gas lighting at the stations. This was removed in the 1930s and it seemed fitting to reinstate this. Lamp columns were sourced from a scrapyard, originally having seen service in the City of Bristol. These were restored and erected at the stations. Lamp heads, faithful reproductions of the originals were supplied by Suggs & Co who had also supplied the original installations over a hundred years ago. Most of the fixtures now use LED technology.

Rheidol Falls displays the results of the restoration project to return stations to an appearance reminiscent of their 1920s / 1930s GWR styling.

THE CHANGING FACE OF PRINCE OF WALES

As related in the text, VoR No. 9 PRINCE OF WALES was subjected to a repaint in 2014, away from the maroon livery first applied following its overhaul at the Brecon Mountain Railway (*upper*) and into Cambrian "invisible green" (*centre*). In the bottom picture, it is seen alongside Great Western liveried No. 8 at Devil's Bridge on 21st September 2014. The two locomotives had arrived on a rare double-headed train.

The evidence of the lineside tree clearances is apparent in the view above of No. 8 heading up Cwm Rheidol, whilst its passengers enjoy the newly-exposed vistas such as that on the left.

In recent years the line has worked tirelessly to control the tree growth along the route opening up once hidden vistas and safeguarding the views for the next generation of railway travellers.

Driver Pete Smith has recorded approximately 150,000 miles on the line completing on average 250 trips a year for the last 25 years, mainly on the regulator of No 9 PRINCE OF WALES. This is the equivalent distance of half way to the moon and the railway believes that Pete may well hold a record for the longest distance travelled by a driver on any tourist railway in the UK. Can anyone challenge this? Pete's contribution was marked with a surprise medal presentation in front of friends and colleagues in December 2014.

Engineering facilities at the railway were for many years very basic with only minor running repairs taking place in Aberystwyth with locomotives and carriages being sent away to Swindon or Oswestry Works for heavy engineering work. This was to change in 2014 when a new engineering workshop facility opened on site at

Aberystwyth. The new workshop was constructed over a period of several years and includes all the facilities required for steam locomotive maintenance including an overhead travelling crane, large lathes, borers and mills as well as an erecting area and inspection pits. A dedicated paint shop and a separate carriage joinery area will ensure that high standards of maintenance and a mirror-like paint finish can be achieved. A viewing gallery has been constructed above the workshop floor and it is intended to allow public access to this at a future date. The winter of 2014/5 saw three Rheidol tanks under simultaneous maintenance on adjacent tracks. "We think this is the first time that this has happened," said VoR CEO, Rob Gambrill, "previously when British Rail operated the line, locomotives were sent to Swindon Works for repairs as the Vale of Rheidol lacked its facilities for heavy maintenance. We have photographic evidence of two locomotives being in the works together, but never all three. We now have a narrow gauge version of the famous 'A Shop' which is able to restore not only our own locomotives and rolling stock, but also take on outside contract work. The three locomotives are lined up on adjacent roads giving an interesting perspective. The facilities at Aberystwyth now allow the railway to take on external projects and contract restoration jobs, a valuable income source for the railway and providing a service to the industry as a whole.

The overhaul of locomotive No 7 OWAIN GLYNDŴR is likely to take approximately two years. So far repairs have taken place on the frames, the cylinders have been removed for re-boring and the frames have been shot blasted and painted. The full overhaul programme is currently being drawn up. This will involve heavy boiler work, new tanks, platework and a full overhaul of the chassis. As much as possible will be done in house by the railway engineering team. No. 7's return to steam is eagerly

In addition to maintenance activities the VoR, like many heritage railways, operates Santa Specials in the winter months. Despite the snow on the ground however, this view of No. 8 is of a spring service on 1st April 2010.

An activity introduced in 2014 are "Driver Experience" opportunities using the line's "Wren ckass Kerr Stuart loco (3114/1918).

awaited having not worked since 1998. All the wheels have been turned, boiler tubes removed and removal of the inner firebox is progressing. No 7 holds the accolade of being the very last steam locomotive to be operated by British Rail when it hauled the final trains of the 1988 season, bringing down the curtain on 40 years of British Rail / state ownership of the line.

Recently a full set of carriages has been turned out in the ornate Great Western Railway chocolate and cream livery complete with the garter crests and gold leaf lettering whilst others are now sporting the GWR shirt button logo, a nod to the railway's heritage. In 2014, locomotive No 9 PRINCE OF WALES was outshopped in Cambrian Railways invisible green. Although not strictly authentic for the locomotive, it was a fitting tribute to the Cambrian Railways in their 150th year.

Also in the works is a historic railway Cattle Van which has recently returned to its former home in Aberystwyth after an absence of over 75 years. Built as one of a pair by the Great Western Railway for transporting livestock, the wagon was supplied new to the Vale of Rheidol Railway to cater for the projected growth in agricultural traffic between Aberystwyth and Devil's Bridge. Sadly this development never materialised with the market being cornered by the growth in popularity of the motor vehicle.

In 1937 the two cattle vans (numbers 38088 and 38089) were transferred to the nearby Welshpool and Llanfair Railway, which was also part of the Great Western Railway. There they saw service until the closure of that line in 1956. In 1960 one of the wagons (38089) was sold on to the Ffestiniog Railway in Porthmadog where it was heavily modified by the Ffestiniog Railway Society East Anglian Group between 1965 and 1968 and saw a variety of uses including as a generator van, a bicycle wagon and more recently as a store room.

One of the demonstration goods trains on which the restored Cattle Van is likely to be used - No. 8 with a rake of wagons alongside the newly restored station at Capel Bangor in October 2013.

In late 2014, the Vale of Rheidol approached the Ffestiniog and made arrangements to re-acquire the van. "This is a part of our history and we thought we should bring it home" said Railway Manager Llyr ap Iolo, "we are dedicated to preserving the heritage of the railway and this was a fantastic opportunity to fill in one of the missing chapters in the history.

The Cattle Van will be restored in our new workshop at Aberystwyth and once completed will take pride of place in our collection of historic wagons. We doubt we'll ever need to use it again for transporting sheep but it will be used for demonstration goods trains and photographic specials and allow us to recreate scenes of a bygone era. Sister wagon, No 38088 has been fully restored at the Welshpool and Llanfair Railway.

In the future the railway plans to build a railway museum on the site at Aberystwyth. This will house locomotives and railway artefacts from the railway's collection, the majority of which have never been on public display before. It is very much a long term project but it is envisaged to provide a world class railway museum.

In 2013 it was announced the company had secured the Victorian roof from London Bridge station which was being removed by Network Rail as part of a major redevelopment scheme at the London terminal. The roof is destined to form the centrepiece of the museum and will certainly make for an impressive setting for the displays.

The Vale of Rheidol runs services most of the year with a daily service between April and October and a more intensive service during July and August. It is not just a destination for railway enthusiasts as people young and old can marvel at the beauty of the Rheidol Valley on one of the most spectacular railway journeys in Britain.

THE SNOWDONIAN
Photographs by Michael Chapman

April 18th 2015 saw this year's running of *The Snowdonian*, one of a small number of trains in the year scheduled to operate through from Blaenau Ffestiniog to Caernarfon. This year's running was blessed with excellent weather, and so we go on a photographic journey behind PRINCE, BLANCHE and LINDA as they head up the Welsh Highland section from Porthmadog to Caernarfon.

Leaving Porthmadog, heading for the new bypass...

...through the Aberglaslyn Pass...

...a breather at Beddgelert...

...then on through the forests...

...over the summit at Pitts Head...

...and round the bends at Rhyd Ddu...

...to descend to Dinas...

...home to bed at Boston Lodge at the end of the day!

NG PEOPLE
photos by Peter Donovan

There are occasions when the volunteers to be seen operating and maintaining our narrow gauge railways create some of the most worthwhile photographic opportunities! Here we see (*above*) the Driver oiling around 822 THE EARL during a layover at Raven Square, Welshpool, on 5th August 2014, and (*left*) an interesting character entertaining the passers-by (and no doubt being cheeky to the photographer!) at Page's Park Station on the Leighton Buzzard Railway.